WEST SCHOOL
1010 FORESTWAY
GLENCOE, IL 60022

$9.95

Australia's
ANIMALS

Australian fur seals, *Arctocephalus pusillus* (see page 41).

Australia's
ANIMALS

Male red kangaroo, *Macropus rufus* (see pages 18 and 19).
(Left) The caterpillar of the emperor gum moth, *Antheraea eucalypti* (see page 59).

Photography by Ken Stepnell
Text by Dalys Newman

CHILD & ASSOCIATES
AN ALL-AUSTRALIAN PUBLISHER

Acknowledgements

The author gratefully acknowledges the following people for their assistance in identification of species:

Taronga Park Zoo Reptile Department.
Geoff Ross, Divisional Supervisor, Australian Division, Taronga Park Zoo.
John Riley, BSC, entomologist.

Taipan, *Oxyuranus scutellatus* (see page 45).

Front cover: Young swamp wallabies, *Wallabia bicolor* (see page 24).

Back cover: Dingo, *Canis familiaris* (see page 38).

Published by Child & Associates Publishing Pty Ltd,
5 Skyline Place, Frenchs Forest, NSW, Australia, 2086
Telephone (02) 975 1700, facsimile (02) 975 1711
A wholly owned Australian publishing company
This book has been edited, designed and typeset
in Australia by the Publisher
First edition 1988
Reprinted January 1989
Reprinted October 1989
© Photographs Ken Stepnell 1988
© Text Dalys Newman 1988
Printed in Hong Kong by Everbest Printing Co. Ltd
Typesetting processed by Deblaere Typesetting Pty Ltd

All rights reserved. No part of this publication may be reproduced, stored in a retrieval system, or transmitted in any form or by any means, electronic, mechanical, photocopying, recording, or otherwise without the prior permission in writing of the Publisher.

National Library of Australia Cataloguing-in-Publication data

Stepnell, Kenneth, 1931-
  Australia's animals.

  Includes index.
  ISBN 0 86777 174 7.

  1. Zoology — Australia. I. Newman, Dalys. II. Title.

591.994

AUSTRALIAN ANIMALS

MAMMALS

Australia is unique in its animal life, truly distinct from the rest of the world. The age of mammals had only just begun some fifty million years ago when the island continent was cut adrift from the surrounding landmasses. The few early groups of primitive mammals already established on the continent—forerunners of Australia's present marsupials and monotremes—found themselves unchallenged by the more advanced types of placental mammals that were appearing on the other continents. And so they thrived in the peaceful environment, taking over from the declining great reptiles and adapting and establishing themselves in every ecological niche.

Marsupials eventually all but vanished from the rest of the world, being replaced by the placental mammals, but Australia remained the 'land of the living fossils'—a marsupial-dominated sanctuary. Though Australia is the acknowledged domain of marsupials, fossils are recorded elsewhere in the world and living representatives still exist, to a small degree, in Central and South America. It is therefore believed that the ancestors of Australia's marsupials originated in South America and arrived in this continent via the Antarctic when the three continents lay much closer to each other.

Australia's early marsupials spread across the continent, inhabiting vastly different regions. In time, they became separated from others of their own kind, and each isolated population developed unique characteristics to suit their specialised environment, eventually varying enough to become a distinctly new species.

And then, no more than twenty million years ago, the marsupials were joined by the first trickle of placental mammals—mainly smaller animals such as the hardy seafaring rats and mice, bats and flying foxes who flew in under their own power, and roving seals who colonised the coastline. Man was the next inhabitant of the continent, perhaps fifty thousand years ago, and with him came the dingo.

A family of grey kangaroos at Zumsteins in the Grampians, Victoria.

The peaks of the Stirling Range rise abruptly above the surrounding countryside in Western Australia. The Stirling Range National Park provides sanctuary for a variety of fauna including western grey kangaroos, brush wallabies, quokkas, honey possums and pygmy possums.

From the first migratory placental mammals evolved a great number of exclusively Australian species. By the time European men made their first landfall there were about 120 species of marsupials, 108 species of native placental mammals and 2 species of monotremes.

Mammals (warm-blooded furred animals which suckle their young) are placed into three groups according to the method of birth. Placentals (e.g., cats, horses, humans) bear their young in an advanced state of development, having obtained nourishment in the womb from the mother's bloodstream through the placental connection of which the umbilical cord is part.

Marsupials (e.g., kangaroos, koalas), having no placenta to nourish the young while in the womb, give birth to sightless, tiny, almost embryonic young who struggle through the mother's fur to the pouch where they find a teat, become firmly attached and remain for a long period of time. Their journey to the pouch is aided by precociously developed forelimbs, armed with sharp, curved claws. The hind limbs, however, are still in an embryonic form of fan-shaped buds. Newborn marsupials are relatively much smaller than those of placental mammals; in marsupials the size of a mouse, newborns are no larger than a grain of rice.

Monotremes (platypuses and echidnas) lay eggs, brood their young in folds of skin and, having no nipples, feed them milk oozing from enlarged pores.

Marsupials
The most outstanding group of Australia's mammals, the marsupials occupy almost every type of habitat: open plains, forests, treetops, even holes in the ground. Through the evolutionary process of adaptation to their environment many marsupials have come to resemble, to a striking degree, those higher placental mammals which occupy the same ecological niche. For instance, there are marsupial mice and blind, burrowing marsupial moles, a spiny anteater that rolls itself into a defensive prickly ball like a hedgehog, possums that swing from prehensile 'monkey' tails and the duck-billed water-dwelling platypus. Quolls or tiger cats occupy a niche that in other continents is filled by the placental cats, and the grazing

kangaroo population fills that of sheep and cattle.

Marsupials can be divided into three groups according to their diet: insect and flesh eaters; grass eaters; and insect, flesh and grass eaters.

The first group, the carnivores, are widely distributed but comparatively less often seen than other groups of marsupials. Most of them are nocturnal, such as antechinuses, phascogales, planigales and dunnarts (types of marsupial mice) which feed on lizards, insects and other small prey. Many of these small insectivorous marsupials are easily mistaken for rats or mice, but any resemblance is superficial. They have sharp-pointed faces compared with the blunt-nosed faces of the rodents and they have white, needle-pointed cat-like teeth instead of the yellowish chisel-shaped teeth of placental rats and mice. They also carry their young in a shallow pouch.

Following the recent extinction of the thylacine (popularly known as the Tasmanian tiger), which resembled a wolf, the largest carnivore is now the Tasmanian devil, a sturdy animal with powerful jaws which make short work of the carrion it feeds on. The largest carnivore on the mainland of Australia is the tiger cat. There are several species and all are spotted, have a bushy tail and display an alert, bold, fearless manner that makes them dedicated hunters. They are ground-dwellers who also climb trees and, unlike most other marsupials, they do not have well-developed pouches, having no more than a shallow furry depression that out of the breeding season is hardly discernible. Another carnivore is the numbat which feeds largely on termites, probing them out of their narrow galleries with its long ribbon-like extensile tongue.

Typically, marsupial carnivores have elongated pointed snouts and the front and back legs are of approximately equal length; the first toe of the hind foot (if present) is at a right angle to the other four. Long and flexible, but not prehensile, the tail may be covered with long or short hair or have a terminal brush. Settlement and development have greatly reduced the numbers of Australia's carnivores.

The herbivores comprise the largest and most well-known group of marsupials, including the wombats, koalas, kangaroos and most possum species. All have well-developed pouches in which the mother carries the young. Koalas and wombats make up a separate branch of this group and appear to be closely related as they share anatomical similarities. The koala differs from the other tree-dwelling marsupial, the possum, in having cheek pouches which are associated with chewing bulky, leafy food; wombats have similar although rudimentary cheek pouches. Koalas and wombats also have pouches which open backwards; the possum's pouch opens forward. A tail is of great use to arboreal animals. Koalas, although now highly specialised for life in the treetops, have at some stage of evolution lost their tail—perhaps indicative of a period of terrestrial life somewhere in their ancestry. The wombat and koala are unique when compared with most other herbivorous marsupials. This is considered to be due to the existence of their common ancestor long before kangaroos and possums, thus giving them a longer period to evolve away from the 'standard model'.

Possums are perhaps the most numerous of the marsupials. Closely related, each of the twenty-four species shares certain adaptations for an arboreal way of life. Each has a long tail, either prehensile (able to be curled around branches) or bushy, fluffy or feather-like and used as a balance and rudder during long leaps or glides. Their most distinctive common feature is the hand-like shape of the hind feet—the clawless big toe is opposable to the rest of the foot, enabling a firm grip on the branches. All the other toes on the hands and feet have strong claws and the second and third toes, like those of kangaroos, are combined. They range in size from the tiny honey possum of Western Australia, which feeds on the nectar of flowers, to the large spotted cuscus of northern Queensland's rainforests, whose diet includes small mammals and birds as well as leaves and fruit.

Grand old gum trees—haunt of Australia's most loved marsupial, the koala. More than five hundred species of eucalypts exist in Australia, distributed over the entire continent.

The macropods are the best known of Australia's marsupials. This group includes the red and grey kangaroos, wallaroos or euros, wallabies, rock-wallabies, tree-kangaroos, quokkas, pademelons, rat-kangaroos, potoroos and bettongs. Most macropods use their hind legs to hop at high speed, both legs moving together. Their long, heavy tail serves as a balance when they are moving at speed or as an extra leg when standing or moving very slowly. All female macropods have well-developed pouches enclosing four teats, and usually bear only one young at a time.

The kangaroos together with the wallabies belong to the Macropodidae family. The name refers to the very large hind feet. The kangaroos include the red of the inland plains, the greys of southern forests and woodlands, and the wallaroos or euros which inhabit rocky hillsides in both coastal and inland areas. The wallaroos are shorter and more heavily built than the reds and greys. There are many species of wallabies and they inhabit more densely vegetated country than the kangaroos. Rock-wallabies are another specialised species of macropod, having developed a highly efficient way of moving with agility and speed in their habitat of rough ranges and rocky outcrops. Pademelons are very small wallabies, with thicker tails, which favour thickets and undergrowth. Another unusual macropod is the tree-kangaroo which leads an arboreal existence despite the rather stiff tail inherited from its hopping ancestors. Quokkas also belong to this family.

The other family of macropods, the Potoroidae are omnivores. They have retained more possum-like characteristics than the Macropodidae. The hind limbs and feet are proportionately shorter than those of the macropods and the tail is prehensile. They are nest-builders and feed on bulbs, tubers, fungi and invertebrates. Within the potoroids is the musky rat-kangaroo, unique in that it uses all four limbs in a bounding gait and usually gives birth to two young. The other potoroids are the delicately built potoroos who are forest-dwellers, and the short-snouted, broad-faced bettongs who inhabit more arid vegetation.

Omnivores are mixed feeders. Bandicoots and bilbies also belong to this group. All the omnivores have long pointed snouts, short powerful forelimbs armed with strong digging claws, and long muscular hind limbs used to propel the animals in a bounding gait. The typical bandicoot makes rough nests and feeds on insects, other invertebrates and the succulent parts of plants. Bilbies include small vertebrates in their diet and make their nests at the end of long burrows. These ground-dwelling animals inhabit all areas of Australia, from the suburbs of the largest cities to the arid desert. Here the attractive rabbit-eared bilby lives in deep burrows to escape the heat and dryness of the desert daylight hours.

The creatures of the arid regions are highly specialised to enable them to survive the extreme heat and to conserve water. Some have no sweat glands. Most have extremely efficient kidneys and excrete greatly concentrated urine. Some marsupial mice can exist without drinking, obtaining all necessary moisture from the insects and reptiles which comprise their diet. The majority of the inhabitants of the desert region, however, are like the bilby and avoid heat stress and water loss by living in deep burrows, often in a state of torpor, similar to hibernation, permitting the body temperature to drop to close to that of the surrounding air.

Other marsupials once inhabited Australia. They are now extinct, though fossils have been found. They were larger than today's species and included giant kangaroos, a wombat the size of a small pony, a large lion-like animal called *Thylacoleo*—a tusked and possibly flesh-eating creature—and a group of plant-eating giants called diprotodons.

Placentals

Marsupials are such a spectacular feature of Australia's wildlife that the native placental mammals are often overlooked. The prehistoric separation of Australia from other landmasses, which favoured the rich development of marsupial life, also resulted in the exclusion of an extremely large variety of non-marsupial mammals, such as the hoofed herbivores, larger rodents and clawed carnivores. However, a great many small non-marsupials have developed within Australia from primitive migrants, evolving into many species which are unique to the country. Most people do not realise that there are almost as many native placentals as marsupials. Our placental mammals include animals as large as the sea lion and as small as the native mice.

Bats are among the most numerous of Australia's mammals. They are the only mammals that fly, propelling themselves through the air by means of wings of thin skin. Each wing extends from the side of the body, backwards from the thumb, between the elongated fingers and thence to the ankle. The wings are really modified forelimbs, long slender fingers like umbrella ribs supporting delicate flight membranes. There is usually also a web between the hind limbs and extending along the tail. Most of the Australian bats are small insectivorous species weighing between 5 and 50 grams, but some fruit-bats have a wingspan of a metre or so and a weight of around 1.5 kilograms. The majority of bats live in tropical and subtropical regions. The fifty or so species include ghost bats, long-eared bats, horseshoe bats, mastiff bats, sheath-tailed bats, bent-wing bats and flying foxes.

Some sixty species of rodents are native to Australia.

(Left)
Euros, red kangaroos and yellow-footed rock-wallabies can often be seen in the rugged hills and gorges of the Flinders Ranges in South Australia—one of the oldest landscapes on earth.

Although Europeans brought the plague-carrying 'ship' and sewer rats and the house mouse to the country, there were many species of rats and mice long before this. It is generally assumed that the ancestors of these rodents were accidentally introduced by means of floating logs and debris, or native transport across Torres Strait. The adaptive force of evolution has been strong in the development of these rodents. Some, like the native hopping-mice, have adapted so well to the central deserts that they do not need to drink. These mice have paralleled the development of foreign jerboas with their large ears, kangaroo-like feet and long slender flexible tails. Most interesting of all the rodents is probably the water-rat, which has joined the platypus in the rivers and swamps and now has partially webbed feet. Some of the native mice and rats have been so long isolated on this continent that they are referred to as the 'old endemics'. These include the hopping-mice, tree-rats, stick-nest rats, rock-rats and the small native mice of the genus *Pseudomys*.

The most recent of the 'native' placental mammals to reach Australia was the dingo. Brought here as a domestic animal by the Aborigines, it has since become feral. The dingo's arrival was so comparatively recent that it can hardly be regarded as a true native animal.

Aquatic mammals who visit Australia's shores include the dugong, or sea cow, and most of the Antarctic seals. The single species of dugong inhabits the northern coast of Australia. Herbivorous animals, they graze on seagrasses in shallow rivers or coastal waters. Like the whale, dugongs have lost their hind limbs, having paddle-like forelimbs and a horizontal tail fluke. Three species of seal breed on the beaches of the mainland and the offshore islands: the sea lion, Australian fur seal and New Zealand fur seal. Other Antarctic seals which can occasionally be seen stranded on beaches are the southern elephant seal, leopard seal, Weddell seal and crab-eater seal. The southern elephant seal used to breed in Bass Strait, but this population is now extinct.

European man brought the remainder of Australia's placental mammals to the country. With him came foxes, rabbits, more rats and mice, deer, water buffaloes, camels and farm animals.

Monotremes

Monotremes are egg-laying mammals, curious paradoxical creatures, which, though typically mammalian in so many characteristics, still retain a number of features from their reptilian ancestors. Among these are their skeletal structure, reptilian egg-laying way of reproduction, reptilian eye structure and only one external opening for both reproduction and excretion (the meaning of the word monotreme). Mammalian features include being warm-blooded, having a covering of hair and suckling their young on milk, the young sucking up the milk as it exudes from the pore-like ducts of the mammary glands.

The world's only monotremes—the most primitive of all mammals—are endemic to the Australian region. The

(Right)
Rainforest—a secluded habitat that is abundant with animal life. Small marsupials, lizards, frogs and colourful butterflies make their home in the tropical, subtropical and temperate rainforests of Australia.

platypus is confined to eastern Australia, the Australian echidna occurs throughout the continent and the long-beaked echidna inhabits New Guinea. Their evolution remains somewhat of a mystery. When the age of reptiles drew to a close about sixty-five million years ago, the mammals began their great period of expansion—a period which continues today. Somewhere along the line the monotremes branched off from the mainstream, or perhaps evolved separately and were bypassed by further evolution.

Despite being the most primitive of all mammals they are nevertheless highly evolved and specialised animals, well equipped to make the best of their environment. The duck-like bill of the platypus is very efficient in sifting small animals such as crustaceans and molluscs from muddy creek beds, and the echidna has a long lightning-fast tongue which is equally efficient at food gathering.

Both these Australian curiosities have been able to ensure their own survival. The platypus's aquatic mobility helps protect it from land-based predators and its chosen habitat—mountain streams and secluded creeks—is relatively free of interference from man. It used to be hunted for its fur but is now protected. The echidna's armoury of sharp spikes, coupled with its remarkable digging ability, makes it immune to most of the predators existing in Australia. Echidnas are capable of clinging tightly to the ground and digging downward into the earth—an excellent way of avoiding enemies as the spines provide a rearguard defence. Greatly enlarged claws and the extraordinary rotary motion of the powerful limbs make this digging action possible. Having survived all the evolutionary hurdles of climatic change and competition from other animals, the platypus and echidna seem likely to continue their unique lifestyle without threat of extinction.

Australia and New Guinea are the only places in the world where the three mammal groups—monotremes, marsupials and placentals—exist side by side. Elsewhere, the placentals assumed an early dominance. In Australia, however, the marsupial's manner of reproduction was a major advantage in its success. By a process known as delayed implantation, some marsupials counter environmental threats to the species more successfully than the placentals. For example, while the tiny baby kangaroo is immature, its mother carries a replacement in her uterus, a fertilised egg held in a state of suspended development while the pouch is occupied. Should she lose the joey, another awaits. This factor, along with the lack of early placental predators, has contributed to the rich and diverse marsupial life of Australia—a truly unique fauna.

REPTILES AND AMPHIBIANS

The first amphibians evolved from a group of fishes about 270 million years ago, becoming the first vertebrates to colonise the Earth's surfaces. Reptiles evolved from a group of early amphibians and dominated the world's seas, skies and land with an astonishing variety of forms during the Mesozoic period, about 120 million years ago.

Today's representatives of these animals are a fairly inferior lot compared with their massive and often fearsome ancestors: amphibians 4.5 metres long and reptiles such as the *brontosaurus* which reached lengths of 26 metres.

Australia today has about 180 species of frogs (the other two groups of amphibians, the worm-like caecilians and the newts and salamanders, are not found in this country) and about 600 species of reptiles. These are approximate figures; new species are still being discovered and known ones reclassified.

Frogs

Australia's frogs can be divided into five families: Myobatrachidae, Hylidae, Microhylidae, Ranidae and Bufonidae. Only one species of Bufonidae exists in this country, the cane toad. It was introduced to Queensland from Hawaii in 1935 and has since spread to New South Wales.

The true frogs (Ranidae), despite being the most abundant frog family in the world, have only one Australian representative—the terrestrial, semi-aquatic wood frog which is found only in the far north of Queensland.

The tree frogs (Hylidae) range from arboreal species with suction pads on their fingers and toes, to terrestrial creatures with larger limbs. They are found in moist coastal and mountainous regions of the north and east of the continent. The largest group, the southern frogs (Myobatrachidae), includes some one hundred burrowing and terrestrial species throughout the country. They have great diversity of habitat, adapting to life in arid regions and alpine bogs.

The Microhylidae, narrow-mouthed frogs, are a small species restricted to the far north of Queensland and the Northern Territory. They have unwebbed fingers and toes.

The frogs found in Australia illustrate an incredible diversity of colouration and form, and possess many specialised adaptations to different habitats. For example, the water-holding frogs which live in the desert regions actually carry their own internal store of water deep into the soil, creating moist surrounds to enable them to survive when the pools dry out.

Crocodiles

Crocodiles are the largest reptiles present in Australia. There are two species—the freshwater crocodile and the saltwater or estuarine crocodile. They both have powerful tails, four-chambered hearts and thick-armoured skin and live in the tropical waters of the north and north-east. Freshwater crocodiles range in length from about 50 centimetres to 2.5 metres, and they have long pointed snouts. They are regarded as more or less harmless, their main food consisting of insects, shrimps, crabs and fish.

Saltwater crocodiles can grow to more than 6 metres in length. These crocodiles are man-eaters. They are extremely aggressive hunters, their usual diet comprising crustaceans, snakes, fish, birds, water-rats and wallabies.

Turtles and Tortoises

The name turtle is restricted to reptiles with paddle-shaped limbs; tortoises are those with walking feet. The large sea turtles of Australia are found in tropical and warm temperate seas and include the loggerhead, flatback, green and hawksbill turtles.

Australia's tortoises are broadly divided into two groups: the long neck or snake neck tortoises and their shorter necked relatives. Snake necks—aggressive carnivores—prefer quiet backwaters, lagoons and billabongs, while the short necks—omnivores—may sometimes be found in swift-flowing streams and rivers. All have the singular habit of pulling their neck in sideways, rather than straight back.

Freshwater crocodiles frequent the lagoons, rivers and billabongs of northern Australia.

Lizards

The necessity of adaptation to a highly variable terrain has produced an amazing variety of lizards, which in diversity of size, form and colour are perhaps unmatched by any other group of Australian fauna.

The skinks are the most numerous family of lizards in the country, with over 150 species being recorded. They vary enormously in size from the land mullet, 75 centimetres long, to creatures a few centimetres in length. Many burrowing forms have suffered complete or partial degeneration of their limbs, causing them to slither along in a manner similar to a snake or worm. A familiar representative of this family is the blue-tongued lizard, often seen in suburban gardens. Another friendly little fellow is the long-tailed spiny skink which can often be seen basking on a stone or old log.

Dragons closely resemble iguanas although they are not related. These lizards are characterised by round heads, stout bodies, powerful limbs, long tails and rough scales. The majority are terrestrial but a large number are skilled climbers and a few are semi-aquatic. Found throughout Australia, many are furnished with an impressive array of spines, such as the thorny devil lizard, or spectacular adornment such as the frill of the frill-necked lizard. The dragons are diurnal creatures, partial to basking in the sun on top of termite mounds or fence posts.

Australia is the home of the monitors or goannas—greater numbers of this species are found here than in any other country in the world. They are Australia's largest

lizards, the biggest of all being the perentie which grows to over 2 metres. Goannas are distinguished from other lizards by their loose rough skin, long necks and bodies, powerful limbs and flicking forked tongues. Bush yarns about the goanna and its habits nearly outnumber those about the kangaroo; few other reptiles have become as prominent in their country's language and folklore.

The pygopods, or legless lizards, are the only family unique to the Australian region, all its members being found within Australia and New Guinea alone. Although the larger species may often be mistaken for snakes, closer inspection reveals blunt, fleshy tongues, visible ear openings, scaly flaps that are the remnants of hind legs, and tails that are easily shed when the pygopod is molested. They are found mainly in the drier regions and the largest member is the common scaly foot, which may measure 75 centimetres in length.

Geckos are found in all parts of Australia except Tasmania and they are all nocturnal. Their soft bodies, lacking the overlapping scales common to most lizards, give them a fragile appearance. Geckos are renowned for their habit of jettisoning their tails during stressful situations. The discarded, wriggling tail is intended to capture the interest of the aggressor while the gecko slithers away. The most commonly encountered gecko is the house gecko which is often seen running around walls and ceilings of houses in northern Australia. Many geckos have adhesive discs on their toes which enable them to move effortlessly on smooth surfaces.

Snakes

Snakes are found throughout all parts of Australia, and about two-thirds of the 160 or so species are venomous. This proportion is much higher than in any of the other continents. Fortunately, less than twenty of these species are actually dangerous to man. All snakes are differentiated from other reptiles by their combined lack of legs, eyelids and ear openings, and by forked tongues which are flicked in and out without the jaws opening.

Five families of land snakes are found in Australia. The elapid snakes, the dominant group, contain all the dangerous snakes, such as the taipan (Australia's deadliest snake), death adder, tiger snake, copperhead, eastern brown snake and mulga snake. The colubrine snakes, despite being the world's largest group, are poorly represented in this country with only ten species, including the non-venomous keelback (a semi-aquatic snake) and two green tree snakes; and the venomous brown tree snake and Macleay's water snake.

The pythons (family Boidae) are all non-venomous and include Australia's largest snakes such as the amethystine python, which has been recorded exceeding 7 metres in length. One of the best known of this species is the carpet python which feeds mainly on bandicoots and other marsupials. These constricting snakes kill their prey by squeezing it in their coils.

Blind snakes (family Typhlopidae) are non-venomous snakes with eyes that have been reduced to small dark marks under the scales. There are about thirty species of these nocturnal snakes that can probably only distinguish between light and dark.

The final group of land snakes belong to the Acrochordidae family, represented in Australia by two non-venomous aquatic snakes found in the extreme north.

Salmon gums with bluebush understorey dominate the woodlands near Norseman in Western Australia. Woodlands are spread across the centre of the continent and comprise some forty per cent of its total area. This attractive habitat is home to some of Australia's most characteristic animals including the larger goannas, agile and whiptail wallabies, brown snakes, shingleback lizards, echidnas, numbats and bettongs.

Rust-coloured termite mounds dot the countryside in the north-west of Australia.

INSECTS AND SPIDERS

Australia's insect life is rich and diverse with many species that are unique in size, colour and profusion. There are at least 54 000 known species of insects in Australia, and probably 100 000 or more species in all.

After the bush flies, cicadas are probably the insect most Australians know well. In summer the bushland vibrates to their song. It is not known how many species there are in the country but the better known include the double drummer, yellow Monday, black prince, fiddler and red-eye.

Australia has the most primitive of the world's species of ant. The larger kinds are known as bull ants, bulldog ants, jumper ants, sergeant ants and inch ants. They may be up to 2.5 centimetres long—hence the name inch ant. Another species of ant, the honeypot, is an excellent example of the ability of some Australian insects to develop resistance to drought. Several members of a colony of honeypots are used as storage casks—using the nectar gathered from glands in the mulga tree underneath which they make their nests. These swollen ants are kept safe from drought in galleries a metre below the surface. They then provide food for the rest of the colony for several months.

Some species of termites take advantage of the temporary abundance of food which suddenly appears in the desert after rain by storing grass in their mounds to carry them over the long droughts. Termite mounds are one of the more conspicuous sights of the Australian outback. They can reach a height of 6 metres. Termites—both the earth-dwellers and the wood-dwellers—are particularly abundant in the warmer regions of the continent.

Nearly all of Australia's butterflies are native, or forms of species occurring in New Guinea and the islands to the north. One exception is the small white cabbage butterfly, accidentally introduced from New Zealand in the 1930s. The huge birdwing butterflies of the rainforests of New Guinea and northern Queensland are among the largest in the world and their wingspan may reach 19 centimetres. Another large butterfly, particularly prized by collectors, is the bright blue and black swallowtail butterfly, the Ulysses, which may grow to a wingspan of 11 centimetres and is found in north-eastern Queensland. The beautiful bright yellow of the *Delias argenthona*, splashed with a band of broad scarlet spots, and the stunning blue, green and purple *Ogyris* of northern Australia are two other species treasured by collectors. Possibly the most common butterfly in Australia is the common grass blue which can be seen in their thousands fluttering through lucerne crops.

Of all the country's insects, the most prolific and most unpopular would be the mosquitoes and flies, the plague of Australia's tropical and subtropical regions. Among the most unpleasant flies in this country are the bush flies which are found in immense numbers in the outback. Research on this species in Canberra some years ago showed that there can be as many as 3600 per hectare.

Australia's indigenous spiders total more than fifteen hundred species. There are several poisonous spiders. The trapdoor spider can give a serious bite but to date has caused no deaths. This spider excavates earth burrows and seals them with close-fitting doors of silk. Related to the trapdoor is the lethal funnel-web spider, which makes a shallow burrow with funnel-shaped silk extensions. Another extremely poisonous spider is the redback, which makes untidy webs under stones, logs and rubbish and underneath houses. A large number of Australia's spiders are orb-weavers, making orb-shaped webs. These include the common garden spiders, tent spiders, golden orb-weavers and Christmas or jewel spiders. Other spiders rely upon hunting and ambush to capture their prey, such as the tiny jumping spiders, the huntsman, crab or flower spiders, and wolf spiders.

One of the most remarkable of all Australia's spiders is the barking spider which has the amazing ability of making a clearly audible barking sound when disturbed. When making this sound they usually rear up on their hind legs. One species of this spider, found in northern Queensland, is Australia's biggest spider, reaching a length of nearly 7.6 centimetres. These spiders have been known to kill newly hatched chickens and drag them down their burrows.

OUR WILDLIFE HERITAGE

Fossil evidence indicates that Australia's fauna has not changed to any great degree since remote times, but wider settlement of the continent is bringing increasing pressure to bear on the wildlife, particularly the native mammals.

The history of settlement has been characterised by the clearing of forests, overgrazing of grasslands and ruthless destruction of any animal that was a threat to the pastoral industry or that could be exploited to become part of the economy. Sealing effectively destroyed the large populations of these mammals around our coasts and it is only recently that seals are reappearing in any numbers. The fur trade took a grim toll on koalas and possums, and to a lesser degree water-rats and platypuses. And the killing of turtles and crocodiles for commercial purposes drastically reduced their numbers.

But man's persecution alone has not been the major cause in the decline of Australia's threatened animals. Most animals are failing in the face of insidious changes in the environment such as clearance. For example, the tropical rainforests of northern Queensland support a specialised and interesting mammal population that is totally dependent on this type of habitat. Should the dwindling remnants of tropical rainforest finally disappear, so too will these mammals.

The noted ornithologist John Gould wrote in 1863:

> Short-sighted indeed are the Anglo-Australians, or they would long ere this have made laws for the preservation of their highly singular and in many cases noble indigenous animals . . . without some such protection the remnant that is left will soon disappear, to be followed by unavailing regret for the apathy with which they had previously been regarded.

This dire prediction has not yet come true. Luckily, despite having been neglected and almost ignored for more than a century, the conservation of wildlife has become an important public issue. It is up to us as individuals to ensure that it remains so: that areas continue to be put aside as wildlife reserves and that specialised habitats are not destroyed; that the comparatively short list of extinct animals be held to its present size and that our remarkable heritage of unique wildlife stays intact.

The Kangaroo Island kangaroo (*Macropus fuliginosus fuliginosus*) is restricted to Kangaroo Island off Yorke Peninsula, South Australia, where it inhabits the dense scrub and bushland, coming out into the open when feeding. This kangaroo is a subspecies of the western grey kangaroo found in the south-western coastal areas of the continent. It differs from other kangaroo species in its dark sooty-brown colouring, heavier build and shaggy coat. A quiet, gentle animal, it is slow moving compared with its mainland brothers, and has shorter ears, muzzle, limbs and tail.

The red kangaroo (*Macropus rufus*), probably the biggest living marsupial, is the most impressive of the great kangaroos. Found throughout the plains and drier inland areas, it has the widest distribution of any kangaroo species. The extreme western and eastern corners of the continent and the mountainous tropical parts of the far northern and eastern coasts are the only areas where the red is not seen. Gregarious in their habits, they are usually found in small mobs of about a dozen or so, but may move in groups of up to two hundred. Although they are wide-ranging, settlement and fencing have restricted their movements and drought proves a serious enemy.

King of the plains—the male red kangaroo. Colouration differs between the sexes of this species. The male, who at about 80 kilograms is more than twice the weight of the female, is a brilliant wine-red colour, and the lightly built doe is a soft-toned smoky blue which has earned her the name 'blue flyer'. The male head *(right)* is characterised by the large and strongly bowed snout which has a well-marked blackish whisker-mark extending towards the eye, accentuated by a whitish area below it. The end of the tail is paler than the body. Red kangaroos are mainly nocturnal, spending their days under shady trees and moving out into the open at night to find food and water. They are also called the plains kangaroo as their long limbs and great mobility are obvious adaptations to grasslands and open plains habitats, although they are equally at home in mulga woodlands and bluebush and saltbush country.

When standing erect, the long and rather slender limbs of the red kangaroo give it an appearance of great height. They are considered to be the largest of the kangaroos, a fully grown male measuring up to 2.4 metres. The usual length of the hop of a red kangaroo travelling slowly is 1.2 to 1.5 metres but with speed this is extended to about 4.2 metres. Their speed is debatable but it is clear from pacing with motor cars, that males can achieve 65 kilometres per hour.

The blue flyer—a female red kangaroo and her joey. Usually only one young is born, but twins may occur. The young first gets out of the pouch at about six months of age for only a few minutes a day. He entirely vacates the pouch about six weeks later. The mother kangaroo helps the young re-enter the pouch by bending down and spreading her forelegs; the young goes in head-first and turns a complete somersault, bringing his head towards the entrance. At the end of pouch life, the female prevents the joey from entering by grasping it firmly with her forepaws or by moving away as it tries to enter. The young suckles from outside until it is about a year old and continues to associate with the mother for a considerable time after that. Red kangaroos cease to breed during severe drought. When conditions are favourable the build-up of the kangaroo population is aided by the ability of the females to carry a reserve embryo in the uterus at the same time as there is a joey in the pouch. Development of the embryo is suspended while the first joey is in the pouch, but within a day of the joey vacating the pouch the second joey is born. Thus births are spaced as closely as possible.

A forest-dweller, the eastern grey kangaroo (*Macropus giganteus*) is found from south-eastern South Australia to Cape York. Among its distinguishing features are the hairy snout, short woolly fur that is grey to grey-brown with a silvery sheen, becoming almost white on the undersurface, and the grey-brown tail, black at the tip. Their preferred habitat is dry sclerophyll forest, woodland and scrubland with adjacent grassy areas. Males are about twice the weight of females and may reach 2.3 metres in length.

Lunchtime. A grey kangaroo feeds a joey nearly as big as herself. The young kangaroo leaves the pouch at about forty-four weeks but continues suckling from an elongated teat until about eighteen months old. Red kangaroos breed at any time of the year if conditions are suitable, but with the grey kangaroos the breeding is more seasonal. The most common months for births are between September and March. The joey then usually leaves the pouch when grass is in good supply.

Grey kangaroos spend their days asleep under low trees and shrubs, and from late afternoon to early twilight move out into open areas to graze on native grasses and shrubs. Their hopping movement is well known. When moving slowly the kangaroo uses its tail, making a 'fifth foot' to help push it forward as it proceeds on all fours in a type of crawl. When it is moving at speed, the gait is a graceful springy bound. The body is inclined forward and the heavy tail acts as a balance behind as the animal springs forward on the tips of its toes.

The whiptail wallaby (*Macropus parryi*) is commonly found on grassy hillsides in wet and dry sclerophyll forest areas, from northern Queensland to north-eastern New South Wales. A large slender animal, it has an exceptionally long tapering tail which is black at the very tip. There is a distinct white face stripe and dark grey at the base of the ears. Whiptail wallabies are gregarious creatures, moving in groups of about fifty. During the early morning and late afternoon they graze on native grasses, herbs and ferns. They sleep for most of the day and at night in the shelter of a shrub or low tree.

At water areas kangaroos are very alert, pausing frequently to observe their surroundings, and when finished they rapidly bound away. Most come to water around dusk and sometimes they may be seen drinking during the day if the water is near the camp. Popular names for the eastern grey kangaroo are the great grey, probably derived from the fact that this kangaroo was so much larger than any of its greyish wallaby cousins seen around the early Port Jackson settlement, and forester because of its natural habitat—the open forest country of the coastal plains and western slopes.

Commonly seen in the tropical north of Australia, the agile wallaby (*Macropus agilis*) can be identified by its uniformly sandy-brown colour and distinct white cheek and hip stripes. It prefers the tall-grass country of the tropical plains, river flats and savannah woodlands, where it feeds in the late afternoon and night, retreating during the day into denser cover of scrub or gallery forests. One of the largest of the wallabies, it grows to 1.7 metres in length, males being twice as heavy as females.

Essentially a bush-loving and browsing animal, the red-necked wallaby (*Macropus rufogriseus*) haunts the brushes and heath country of the low coastal tablelands and the dense undergrowth of the ranges up to high altitudes. They are commonly found on the eastern and south-eastern mainland and in Tasmania. The large, gracefully built animals vary considerably in colouration but can be distinguished from other wallabies by the contrast of the reddish nape and shoulders with the greyish-fawn colour of the back. They can grow to 1.8 metres in length and the males are noticeably larger than the females. During the day they sleep in dense vegetation; in the late afternoon they graze close to the forest edge, moving into more open areas after dark.

Young swamp wallabies (*Wallabia bicolor*). This species occurs along the eastern coast from Queensland to Victoria and South Australia. Of a large, stocky build (1.4–1.7 metres) they have a dark brown back, contrasting with a rich rusty yellowish belly and black tail. Although at home in swampy or marshy country, the swamp wallaby is also found on scrub-covered hillsides and mountain heights, from tropical to cool-temperate climates. During the daylight hours they sleep in dense vegetation, emerging from the protective cover at dusk to feast upon native shrubs, cultivated pastures and pine seedlings.

Numerous small populations of the parma wallaby (*Macropus parma*), a very rare species, can be found in the relatively undisturbed valleys of the Great Dividing Range. This finely coloured wallaby has a uniformly reddish back and clear white throat contrasting with the dark sides of the neck, a white cheek mark and dark neck stripe. It was thought to be extinct but in 1965 it was rediscovered on Kawau Island, off the coast of New Zealand, where it had been introduced long ago. Soon afterwards, this little wallaby was found to be still surviving in New South Wales. It is one of the smaller wallabies, 97 to 108 centimetres in length, the males being larger than the females.

The brush-tailed rock-wallaby (*Petrogale penicillata*) can be seen on the cliffs and rocky slopes in the mountainous regions of New South Wales. This stoutly built wallaby is a rich dark colour and its distinctive feature is a strongly brushed tail. Rock-wallabies' tails differ from those of most kangaroos in that they are unusually long and slender, and bushy near the end. The tail is held out behind when the animal is leaping over rocks and obviously acts as a balancer. Another unique feature of the rock-wallaby is its feet: they have large thick pads with rough granulations over the soles, similar to the tread on a sandshoe. This extremely agile wallaby sleeps by day in the shelter of rocks and feeds at night on native grasses and other vegetation.

The small tammar wallaby (*Macropus eugenii*) has the distinction of being the first marsupial seen by European man in the Australian region. This was in 1629 on the Abrolhos Islands in Western Australia, when the Dutch ship *Batavia* captained by Francisco Pelsaert ran aground. He described what he saw (tammars) as 'large numbers of cats, which are creatures of miraculous form, as big as a hare'. These wallabies are today found in Western Australia, South Australia and nearby islands, including Kangaroo Island. The small creatures travel almost silently through established runways or paths to graze in more open areas. When resting (pictured) they assume the position of a female kangaroo giving birth—the tail is forward between the legs. In arid coastal regions some tammar wallaby populations are known to drink sea water—the kidneys are most efficient in handling the excretion of salt. An adult tammar stands about 60 centimetres in height. The female weighs about 4 kilograms and the male about 6 kilograms. The back is dark brown with a faint stripe running from the head to the mid-back. They have a white upper lip, throat, chest and belly.

The largest (1.3 metres) and most strikingly coloured of the rock-wallabies, the yellow-footed rock-wallaby (*Petrogale xanthopus*) is found in the arid regions of South Australia and New South Wales. Favourite haunts are rugged rocky ranges with open woodland and acacia scrubland. This wallaby is distinguished from all others by the ring markings on its tail, bright yellow hind feet and forearms, and large furry yellow and white ears. The skull is larger than any of the other rock-wallabies and is characterised by its unusually elongated muzzle and lower jaw. Days are spent asleep among vegetation between boulders or in a rocky cleft and they come out to feed on native grasses in the evening. In winter, rock-wallabies often emerge from their cliffs and caves during the day to sun themselves.

Dense forest country, particularly rainforest, is home to the pademelon. These small marsupials are predominantly browsing animals and their feet and tails are comparatively shorter than those of the typical wallabies and kangaroos. This red-legged pademelon (*Thylogale stigmatica*) is found in rainforest and wet sclerophyll forest in eastern Queensland and New South Wales. It grows to a metre in length and has brownish fur, a yellow hip stripe, rufous heels and an indistinct stripe down the back of the neck. It sleeps from mid-morning to mid-afternoon, supported by a tree or rock, and forages in the evening and night for fallen tree leaves and berries of shrubs and native grasses.

The red-necked pademelon (*Thylogale thetis*) is found in the rainforest and wet forest country from Bundaberg, Queensland, to Moruya, New South Wales. Colouring varies, the deep reddish shoulders being less marked in some localities. The pademelon's habit of living in ferns and other thick vegetation, through which the animals make a maze of tunnels, means that seeing them in the wild is difficult. They have a crouching-hopping motion, well suited to the thick undergrowth. When alarmed they thump out a warning signal with their hind feet—a habit common to most members of the kangaroo family.

A unique member of the wallaby family, the quokka (*Setonix brachyurus*) is found on Rottnest Island off Fremantle in Western Australia, and less plentifully on Bald Island and the coastal south-western areas of the mainland. Apart from its small size, about that of a hare, the quokka is distinguished from all other pademelon wallabies by its short tail, short feet and short rounded ears which can hardly be seen above the long fur. It has a rather shaggy coat and prefers swampy tracts and thickets. The Dutch navigator Willem de Vlamingh named Rottnest Island after the quokka, which he believed to be some species of rat; the word Rottnest meaning rat's nest.

The rare, possibly endangered, brush-tailed bettong (*Bettongia penicillata*) is restricted to small areas in south-western Australia and Queensland. Up to 74 centimetres long, they have short rounded ears, strong front claws and a strong tail which is tufted on the end. During the day bettongs rest in grass nests and at night they go out to forage for succulent roots, fungi, seeds and insects.

The potoroo was one of the first of Australia's native mammals to become known to the world. Governor Phillip described and illustrated in his *Voyage to Botany Bay* in 1789 an animal he called the 'Kanguroo Rat'. It was soon discovered that the Aboriginal word for this animal was potoroo. It is one of the fastest moving of all Australian mammals, moving in an unusual style with its front legs tucked under its chest, its body horizontal and parallel to the ground and its rear legs driving it forward like a bullet. The long-nosed potoroo (*Potorous tridactylus*), pictured, is found in Tasmania and on the south-eastern mainland in wet sclerophyll forest, cool rainforest and heathland. It is 58 to 65 centimetres in length. By day, these animals sleep in a nest of vegetation, coming out at night to dig in the soil for succulent roots, fungi and insect larvae. Potoroos have become rare in most parts of Australia due to the destruction of much of the dense vegetation which is essential for their protection and shelter.

Lumbering and bear-like, the common wombat (*Vombatus ursinus*) is found along coastal forests of the south-eastern mainland and in Tasmania. They are extremely compact animals, weighing up to 35 kilograms but rarely more than a metre long. Enormously powerful forelimbs equipped with impressive claws enable wombats to create extensive burrows. They shelter in these burrows during the day, going forth at night to feed on grass, roots and herbage. Being marsupials, they give birth to a very immature baby which is carried in the pouch for about six months.

The hairy-nosed wombat (*Lasiorhinus latifrons*) is distinguished from the common wombat by its distinctive hairy snout, bigger ears and fine silky fur. They are found on arid coastal and inland plains of Western Australia and South Australia. This species adapts to its arid habitat by lowering its metabolic rate when food is sparse. They do not need to drink and can go without water for three or four months, sleeping in their deep, cool burrows, conserving their energy to avoid loss of water. Burrows are often clustered together to form extensive warrens, with entrances meeting in a large communal pit. Hairy-nosed wombats graze at night on tough native grasses.

(Right)
The koala (*Phascolarctos cinereus*), the best known and most loved of all Australia's native animals. This cuddly-looking animal is found in wet and dry eucalypt forests and woodlands along the eastern coast, extending inland to the western slopes of the Great Dividing Range. Settlement has greatly reduced its range, not only because of the clearing of forests. In the first quarter of this century alone it is estimated that ten million koalas were shot for their pelts. Today they are rigorously protected. Koalas are soft, grizzle-grey furry animals, about 75 centimetres from nose tip to where the tail would be if they had one. They are highly specialised for arboreal life; their hands have a vice-like grip between the first two and other three fingers, and together with the very long arms and curved claws, this enables them to climb smooth tree trunks. The tail is replaced by a callused pad which enables them to sit for long periods in tree forks without discomfort.

Gum leaves are the koala's sole food and there are strong local preferences regarding the species of eucalypt; smooth-barked trees of high oil content, such as the blue and grey gums, are preferred in New South Wales, and the manna gum in Victoria. A kilogram of foliage is the average daily consumption. To accommodate this bulky food the koala's 'appendix' is extraordinarily large—up to 2.5 metres long. Gum leaves seem to supply all the moisture the animal needs, for it rarely drinks water. In fact, the name koala was derived from an Aboriginal word meaning 'no drink'.

The koala is perhaps less nocturnal in habit than many other marsupials. Although it is often seen sleeping or drowsing through the day it is liable to wake up at any time and start feeding on nearby leaves. Towards evening, the animal climbs from its perch, usually in a low fork, up to the higher branches to enjoy a large meal. The slothful appearance of koalas is deceptive: they can move rapidly through the branches and jump for a metre or so from branch to branch.

Even when weaned to a diet of gum leaves, the young koala stays with its mother until quite large. Koalas produce only one young at a time, the baby being extremely small at birth—2 centimetres long—with strong forelimbs to assist it on its way to the mother's pouch. It does not appear outside the pouch until it is about six months old and about 18 centimetres long. It uses the pouch for another two months, afterwards being carried on the mother's back or hugged closely to her when resting or cold. The koala's pouch is unusual in that it opens backwards, differing from possums, the other tree-dwelling marsupials. The young is weaned from milk to gum leaves in a most unusual way. When the baby is about twelve months old the mother begins feeding it pellets of faecal material from her digestive tract.

WEST SCHOOL
1010 FORESTWAY
GLENCOE, IL 60022

Most widespread of the gliders is the sugar glider (*Petaurus breviceps*). It occurs in the wet and dry sclerophyll forests of the northern and eastern mainland, from the Kimberleys to Adelaide, and in Tasmania. It is the smallest glider (32–42 centimetres) and can glide for remarkable distances: 50 metres has been recorded. Gliders leap from tree to tree by means of their flight membranes, an extension of the body skin which is joined to the legs so that it is tightly stretched when the animal leaps from a tree with all four limbs outspread. The tail acts like a rudder. Sugar gliders are very active and fast moving in their nocturnal hunting for insects, blossoms, native fruit, acacia gum and sap which is obtained by biting pieces out of tree trunks. In the daytime they sleep with other family members in leaf-lined nests in tree hollows.

Closely related to the sugar glider, the squirrel glider (*Petaurus norfolcensis*) is larger (43–53 centimetres) and prefers a drier habitat. Its range is less extensive, being recorded from Victoria to north of Cardwell in Queensland. They inhabit coastal ranges, favouring fairly dense white-barked eucalypt country. Their colour differs slightly from the more common sugar glider, the undersurfaces being white to creamy white instead of a very pale grey to medium grey, and the dark dorsal line is usually more distinct. It has a fluffier and more squirrel-like tail which has lovely flounces of grey fur at the base. When annoyed it makes a curious whirring noise and if handled bites savagely. Squirrel gliders are agile climbers and feed on sap, gum, pollen, nectar and insects.

(Right)
Brush-tailed possums (*Trichosurus vulpecula*) are the most common of Australia's marsupials, ranging all over the mainland and in Tasmania. They are more adaptable than most pouched animals, living under many different conditions, sometimes taking up residence in a house roof. Their thick, close, woolly fur is of economic value and they have suffered ruthless exploitation in the past. Fast, agile climbers, they are silvery grey with a long bushy tail, long pointed ears and a rather pointed foxy face. At night they feed on eucalypt leaves, fruit and blossoms, and by day they sleep in a tree-hole or similar cavity. They range in weight from 1.5 to 4.5 kilograms.

The eastern pygmy possum (*Cercartetus nanus*). Although mouse-like in size, these animals differ from mice in that they have prehensile tails and pouches to carry their young. Huge bulging eyes and sensitive ears guide them in their nocturnal hunting for insects, fruit, nectar and pollen which is gathered with a brush-tipped tongue. They sleep by day in woven nests in tree-holes or crevices. As winter approaches they store fat in the base of their tails to tide them through the coldest months when they become torpid. (Torpor resembles hibernation but is a much more temporary state.) A pygmy possum may become torpid for a few hours or days, entering a deep sleep and allowing its body temperature to drop to within a few degrees of its surroundings, thus conserving energy. These small marsupials (14–21 centimetres) are found in temperate rainforests, woodlands and heaths on the south-eastern mainland and in Tasmania.

The southern brown bandicoot (*Isoodon obesulus*) is found in woodland or scrubland areas with low ground cover on the south-eastern mainland. Bandicoots, who are omnivores, have the front teeth of carnivores but the fused rear toes of herbivores—the latter being regarded as a climbing or prehensile adaptation of use in combing the fur which is much infested with lice and ticks. These interesting animals sleep by day under the surface of a well-built nest of earth, grass and twigs, which is sometimes situated in a scrape in the ground or, in damp conditions, may be raised on a platform on the soil. They feed at night, foraging for worms and insect larvae with their front paws. They are fiercely territorial animals and grow to 50 centimetres in length.

Found throughout central and northern South Australia, the crest-tailed marsupial rat or kowari (*Dasyuroides byrnei*) is a desert animal. They live in burrows on the stony and sandy tablelands and prey upon large insects, rodents and other small ground-dwelling vertebrates. They do not need to drink. From 25 to 32 centimetres in length, the kowari has a pointed face, bright eyes, short ears and clawed feet on thin but strong legs. Its distinguishing feature is a long tail with a terminal brush of bristly black hairs.

The isolated highlands of Tasmania are the haunts of the ferocious-looking Tasmanian devil (*Sarcophilus harrisii*). This strange marsupial was one of the first mammals observed by the earliest settlers in Van Diemen's Land and its forbidding appearance earned it the name of 'devil'. The devil's voice probably contributed to the satanic effect for the colonists—it is a whining growl followed by a snarling cough or a low yelling growl when angry. Related to the native cats, this animal is about a metre long and has relatively short legs, a short broad muzzle, small eyes and broad rounded ears. Its thick fur is black or dark brown with white patches. Its favourite food is carrion and when this is lacking it preys on other animals. Three or four young are born in the typically immature marsupial fashion and enter the rearward-facing pouch, where they remain for about fifteen weeks. The devil, despite its appearance and manner, is a shy animal and almost wholly nocturnal in habit.

The sturdy spotted-tailed quoll (*Dasyurops maculatus*) is the largest known marsupial carnivore on the mainland. The adult can measure more than a metre from nose to tail tip. Ferocious and stubborn animals, they are stealthy, solitary creatures of the treetops, with great climbing ability and remarkable hunting skills. Their food consists of any small animals, including rabbits and smaller mammals, birds and their eggs, and reptiles. They sleep most of the day in their nest in a hollow log or rock crevice and feed at night. Found throughout the eastern mainland and Tasmania, their preferred habitat is cool temperate to wet sclerophyll forest and rainforest.

The stripe-faced dunnart (*Sminthopsis macroura*). Relentless hunters of almost anything that moves, dunnarts inhabit heaths and grasslands. Some species have the ability to store fat in the tail—taking advantage of a good season to eat as much as they can in preparation for the possible scarcity of food. They can also enter a state of torpor, similar to hibernation, when food is short, presumably a way of eking out the store of fat in the tail. These delicately built marsupials sleep during the day under cover or in soil cracks, and at night forage for insects and other arthropods. When threatened they open their mouth, bare their efficient array of teeth and utter hissing sounds.

The eastern quoll (*Dasyurus viverrinus*) is much smaller than its relative, the spotted-tailed quoll. Once widespread on the mainland, it now apparently survives only in Tasmania where it is fairly common. It is almost wholly a ground-dweller, sleeping throughout the day in small caves, rock crevices or hollow logs and emerging at dusk to hunt small birds, mice and rats, lizards, insects and young rabbits. They favour wet to dry sclerophyll forest and can sometimes be seen in savannah and heathland. All are spotted on the back but lack spots on the tail. Length: 52–73 centimetres. The native cats differ from most other marsupials in lacking the well-developed pouch of kangaroos, koalas, possums and bandicoots. The cat pouch is usually only a shallow furry depression containing the teats.

Found in wet eucalypt forest with dense ground cover, the brown antechinus (*Antechinus stuartii*) makes its home in a variety of places such as tree hollows and stacks of dead wood. Its nest is made of leaves and bark. These common marsupial mice are mainly nocturnal, their food consisting of insects and other small animals. They sit up in a squirrel-like fashion to devour their prey with needle-sharp teeth. Antechinuses have long claws which make them efficient climbers and they grow to 25 centimetres in length.

Although known to the early colonists as a squirrel because of its appearance and agility in the trees, the brush-tailed phascogale (*Phascogale tapoatafa*) is more like a weasel in the general appearance of its lithe body and its bloodthirsty habits. Insects and small vertebrates are its main food, but it has a reputation for attacking penned poultry. The fur is a beautiful bluish-grey colour and when alert the hairs on the tail are spread out like a bottlebrush. The phascogale spends the day in a nest of bark, leaves or grass in a tree-hole and hunts at night. They are found in open, well-watered sclerophyll forest in coastal areas in the north, south and south-east.

Spinifex hopping-mice (*Notomys alexis*) are a desert species that are well adapted to their environment and can exist without water. They avoid daytime heat, remaining in their complex burrows a metre or more down in the cool and usually damp earth. Emerging at night, they feed on seeds, shoots, roots and insects. Remarkably elongated hind feet distinguish the hopping-mice from all other native rats and mice. They have evolved a two-footed leaping action, similar to that of kangaroos, and their general appearance is that of a miniature kangaroo. The majority of the Australian hopping-mice species inhabit northern and arid regions.

The swamp rat (*Rattus lutreolus*) is found in heathland or grassland and densely vegetated swamps in areas of rainforest or sclerophyll forest. It is a plump animal, 21 to 34 centimetres in length, with a tail which is shorter than the head and body. The fur is dark brown on the back and dark grey below. This species, which is distributed throughout eastern Australia, makes tunnels in dense undergrowth around swamps; they do not seem to mind wet and muddy conditions underfoot. They are strictly vegetarian, feeding mostly on grasses, sedges and seeds. There are two other subspecies: *R. l. lacus* in Queensland, and *R. l. velutinus* in Tasmania.

(Right)
The most recent of 'native' placental mammals to reach Australia was the dingo (*Canis familiaris*), brought here by Aborigines at least five thousand years ago. Found throughout the mainland, but not in Tasmania, dingos favour grasslands and the open plains. They do not bark like an ordinary dog, but make a sustained dismal howl, similar to that of a wolf, and that only at night. Hunting is done at dusk and they remain in hiding during the day. Colour varies but they are usually a tawny yellow with a blunt head, short ears which remain erect and a bushy tail. Five pups form the average litter.

(Left)

The Australian fur seal (*Arctocephalus pusillus*) population was severely reduced in the early nineteenth century when they were ruthlessly killed for fur and oil. Today the population has recovered to about twenty thousand. These seals inhabit cool temperate coastal seas and breeding colonies are found scattered along the south-eastern coast of Australia. Large animals (males may grow to about 2 metres in length), they feed mainly on squid and octopuses, fish and rock-lobsters, and they can dive to at least 150 metres. The colonies are a noisy and colourful place during the summer breeding season. The males establish a territory on the beach, fighting off the younger bulls, and the females give birth to their pups, and then spend endless hours teaching them to swim and catch fish.

Bent-winged bats (*Miniopterus schreibersii*) are perhaps the most numerous and most common of the cave-dwelling bats, particularly in the highlands of eastern Australia. These small brown bats have a wingspan of less than 25 centimetres. They are insect-eaters and feed at night, emerging from their caves to hunt above the tree canopy for flying insects. They roost communally in caves and mines and other man-made constructions, and, after mating, the females make their way to special maternity caves in which they congregate to give birth. Soon after the young are born, the females and their offspring pack together in dense clusters in the caves, forming a living carpet on the walls. They do this to conserve heat and humidity, essential for the development of the young bats. More than 3700 baby bats per square metre have been counted on one cave ceiling.

The spiny anteater or short-beaked echidna (*Tachyglossus aculeatus*) is, along with the platypus, an egg-laying monotreme, the most primitive of mammals. One of the real wonders of the animal world, the echidna lays eggs yet suckles its young; it has no teats but exudes milk through its pores. The female grows a pouch only when required for carrying its young and after this function has been accomplished the pouch closes up. She apparently lays her solitary soft-shelled egg directly into this pouch. There she incubates it until the naked young is hatched and she then broods it until its sharp spines develop. Then it is evicted and hidden in a concealed place. The young is suckled for at least three months. Echidnas are 45 to 50 centimetres long and are covered with short, sharp quills which form a protection against its aggressors. The curious creature has a long tongue covered with a sticky substance, and it obtains its food by thrusting this tongue into ant hills and withdrawing it covered with hundreds of ants. Echidnas are found throughout Australia and their natural habitat is open forests, scrublands and rocky areas; they can often be seen in suburban gardens.

The platypus (*Ornithorhynchus anatinus*) is an extraordinary combination of bird, reptile and mammal. They inhabit river banks and lakes from Tasmania to northern Queensland but are not often seen because of their unobtrusive habits and shy nature. An adult platypus measures about 60 centimetres in length and can weigh up to 2 kilograms. Their thick, soft fur is brown above, shading to creamy grey below. The tail is broad and flattened, similar to that of a beaver, and is covered with fur. The bill is superficially duck-like but very broad and prominent, and covered with thick, soft naked skin which is blue-grey on its upper surface. The adult has no true teeth, horny ridges taking their place. Eyes are small and set in narrow furrows which also contain ear openings. The paws are webbed, the front ones having a further flap of skin extending well beyond the toes. When the platypus is walking or digging, this flap remains tucked under the palm of the hand, exposing the claws. When swimming, however, the platypus unfolds the flap so that it acts like a flipper. The platypus sleeps for most of the day in a burrow in the bank of the river. It feeds at night on a wide variety of invertebrates taken from the bed of its watery habitat, sifting its food from mud and water by means of its flexible snout. Although considered an aquatic creature, the platypus can actually only remain underwater for a few minutes and shuts its eyes and ears before submerging. The retention of the egg-laying habit in this mammal makes it a zoological curiosity: all other living mammals (except echidnas) bear their young alive. The female platypus lays her eggs in a nest in a round chamber at the end of a burrowed tunnel and curls her body around the eggs to incubate them. She feeds her hatched young with milk; she has no teats and the babies obtain milk by sucking it through enlarged pores of her skin.

Most commonly found in the regions around Mount Gambier in south-eastern Australia, the lowland copperhead (*Austrelaps superbus*) can grow to 1.7 metres. It is a slow-moving, heavy-bodied snake and when threatened erects the front third of its body, adopting an aggressive cobra-like pose. Although venomous, the copperhead is usually inoffensive and bites are uncommon. Cold-blooded vertebrates, especially frogs, are its main source of food and it is active both during the day and at night, even at very low temperatures when no other reptiles are active. This low-temperature tolerance is one of the best distinguishing features of these snakes: they are active earlier in spring and later in autumn than most other snakes. Colour is variable, ranging from light grey through chocolate brown to black.

One of Australia's most notorious venomous snakes is the eastern or mainland tiger snake (*Notechis scutatus*). It is found in several environments, from rainforest to dry sclerophyll forest, throughout much of coastal and south-eastern New South Wales, Victoria and south-eastern South Australia. Australia has two types of forest: the rainforest and the eucalypt-dominated sclerophyll forest which is classed as wet or dry depending on situation. The dry sclerophyll forest receives less rain and is on a less fertile site.

The eastern tiger snake's preferred food is frogs and tadpoles and they therefore frequent watercourses, swamps, lakes and wet mountain slopes. Although one of the most common of Australia's snakes, draining of lakes and swamps for farming purposes and control of flood levels along rivers has reduced the tiger snake population. Up to 1.2 metres long, the tiger normally hunts during the day but can sometimes be seen slithering along on warm nights. Like all snakes, it uses its teeth for gripping and biting, gradually working its way along the victim's body. While eating, it breathes through a windpipe located at the side of the mouth.

Streams, swamps and lagoons of eastern Australia and south to south-eastern South Australia are the habitat of the red-bellied black snake (*Pseudechis porphyriacus*). It is a fine-looking snake, irridescent black on top with an intense red on the outer edges of the belly, fading to a lighter colour towards the centre. A voracious eater, it is active during the day, feeding on frogs, small mammals, other snakes, lizards, fish and eels. Although poisonous, the bite of a red-bellied black snake is not life-threatening. Usually shy by nature, the snake flattens out its neck when alarmed and may even attempt mock strikes. The young—numbering up to forty—are born in membranous sacs from which they emerge several minutes after birth.

The water python (*Liasis fuscus*), a uniformly black or brown nocturnal snake, is found in northern Australia. Usually seen near streams, dams, lagoons and billabongs, it takes to water when alarmed. Its average length is 2 metres. This family of snakes (Boidae) includes the largest snake in Australia, the Queensland rock python (*Liasis amethystinus*), which grows to 6 metres in length. Pythons are all non-venomous, egg-laying snakes, many of which are known to incubate their eggs. Most of them feed on warm-blooded animals which they kill by constriction or suffocation.

Found throughout Australia, the common death adder (*Acanthophis antarcticus*) is a secretive nocturnal snake which spends the day half buried in sand, soil or litter, often at the base of trees or shrubs. Its average length is 40 centimetres and it has a short stubby body with a small, thin rat-like tail ending in a curved soft spine. One of the country's most dangerous species, it possesses well-developed fangs and venom glands and strikes with amazing speed from a tense, flattened, coiled position. In the wild, death adders feed on native insectivorous reptiles, birds and mammals, which they capture by wiggling their tails with an insect-like movement to act as a lure.

Australia's deadliest snake, the taipan (*Oxyuranus scutellatus*), is found in Queensland and the northern part of the Northern Territory. It has a wide range of habitats from tropical wet sclerophyll through dry sclerophyll and open savannah woodland. The sugarcane fields of Queensland provide an excellent habitat, and the rats which inhabit these areas are an ideal food source. Up to 2.8 metres in length, taipans can always be identified by their pale, creamy-coloured heads. They are extremely efficient hunters, feeding on rats, small birds, mice, bandicoots and lizards. Of all the Australian snakes, the taipan is the most intelligent, nervous and alert. Because of its keen senses it can usually retreat from humans, but if approached suddenly, or cornered, it viciously defends itself, often striking several times.

Lesueur's gecko (*Oedura lesueurii*) is a small common gecko found throughout southern and eastern Australia. It usually lives in caves or crevices, where it shelters under stones and feeds on small insects on the open rock faces. Like all geckos it is able to cast off its tail when attacked. The violent contractions of the dismembered tail attract the attention of the predator, enabling the gecko to escape. The tail grows back within a short time. Lesueur's gecko grows to a length of 10 centimetres and is a pale bluish-grey colour with a distinctive pattern of squarish blotches.

The strange-looking knob-tailed gecko (*Nephrurus levis*) has a knob on the tip of its fat tail and is common in arid areas of Australia from the central coast of Western Australia to the arid parts of all mainland States except Victoria. It is a nocturnal terrestrial lizard which often shelters in burrows in the sand during the day and comes out at night to forage in open areas. It can reach up to 8 centimetres in length.

The semi-arboreal bearded dragon (*Amphibolurus barbatus*) is often seen sitting on fence posts, basking in the sun. This lizard is widespread in the eastern half of Australia and the far north, favouring habitats as varied as dry sclerophyll forest and desert country. Camouflage is its principal form of protection but if molested it will resort to a spectacular defensive stance *(below right)*. It opens its mouth fully to display the bright yellow interior, the magnificent 'beard' is extended and the ribs expand, giving the body a greatly enlarged appearance. The bearded dragon can grow to 75 centimetres and it feeds on insects, worms, mice, small lizards and soft ground herbage. An egg-layer, it deposits up to twenty-five eggs in a shallow hole scooped in sandy soil.

The fast-moving tawny dragon (*Amphibolurus decresii*) is found in rocky areas in South Australia, including islands off the coast. They are usually a greyish-brown colour with the male often being identified by cream, yellow or orange spots on the side of the neck. The tawny dragon grows to a length of 25 centimetres and feeds on small insects including moths, beetles and grasshoppers.

The semi-aquatic eastern water dragon (*Physignathus lesueurii*) is found living beside waterways along the eastern coast, between northern Victoria and Cape York. An arboreal lizard, it is often seen perched on tree branches overhanging the water in which it takes refuge when threatened. Male water dragons grow larger than females, specimens of 90 centimetres having been recorded. They feed on insects and aquatic organisms including frogs, as well as other small terrestrial vertebrates, fruits and berries. A very able swimmer, the water dragon propels itself through the water by means of its strong tail.

The swift-moving painted dragon (*Amphibolurus pictus*) is a diurnal lizard found in the drier parts of southern Western Australia through South Australia to north-western Victoria and central western New South Wales. During the mating season the male painted dragon is undoubtedly Australia's most colourful lizard, acquiring a bright blue flush on the throat and flanks and bright yellow or orange on the chest. They grow to about 25 centimetres and ocupy shallow burrows in sandy soils, usually at the base of saltbush or other low scrub. They feed on small insects and when agitated the male has the ability to raise his dorsal fin.

The lace monitor (*Varanus varius*) is Australia's second largest lizard, growing to a length of 2 metres. (The perentie is the largest.) These creatures are arboreal and may take to the trees when startled. They are very capable hunters, feeding on insects, reptiles, small mammals and nestling birds. They also feed on carrion and many lace monitors may be seen feeding off the same carcass. Regions they are found in include the coast, ranges, slopes and adjacent plains from south-eastern South Australia to Cape York. Females lay their eggs inside termite mounds, and on hatching, the young are strikingly banded with yellow and black and measure about 28 centimetres.

At home in the water and on land, Merten's water monitor (*Varanus mertensi*) is found near the permanent waterways of central Queensland, the coastal Northern Territory and the Kimberley region of Western Australia. It is a fairly common aquatic lizard and is often seen basking in the sun in the early morning and late afternoon. Most of its food—crustaceans, fish, large water insects and frogs—is found beneath the water where it walks on the river bottom with its eyes open. When alarmed, its powerful vertically compressed tail propels it swiftly through the water with speed equal to that of a crocodile. A solid creature, it reaches a length of 1.25 metres.

Despite its fearsome looks, the slow-moving thorny devil (*Moloch horridus*) is completely inoffensive, apparently relying upon its grotesque appearance to ward off predators. When attacked it will retract its head under its body and present its enemies with the spiky hump on its neck. This unique lizard inhabits the desert and semi-desert areas of central and south-western Australia and grows to 15 centimetres in length. Well adapted to arid environments, the thorny devil can soak up water with its skin. It feeds only on ants and may eat thousands during the course of one meal.

Often encountered in suburban gardens, the eastern blue-tongued lizard (*Tiliqua scincoides*) is a diurnal ground-dweller, feeding on a variety of insects, snails, carrion, wildflowers, native fruit and berries. It shelters at night in hollow logs and ground debris. If in danger the lizard usually retreats, but if cornered it will hiss violently, opening its mouth and producing its bright blue tongue in an attempt to frighten its aggressor. It is found in a wide variety of habitats in south-eastern South Australia, through Victoria, eastern New South Wales and most of Queensland to the northern Northern Territory and north-western Western Australia. It reaches a length of 50 centimetres.

The well-known shingleback lizard (*Trachydosaurus rugosus*) is widely distributed in semi-deserts, savannah and open forests throughout the southern half of Australia. Its colour is variable, but it is usually dull reddish brown, dark brown or black with scattered cream or yellow splotches. Grossly enlarged scales, like a pine cone in appearance, cover its tail and upper body. A slow-moving diurnal, it grows to 40 centimetres long and feeds on insects, snails, carrion, flowers, fruit and berries. Although fairly inoffensive, when cornered it adopts a U-shaped stance, aggressively flicks its tongue and may inflict a painful bite if sufficiently provoked. It is preyed upon by feral cats and foxes.

The legless lizard (*Delma* sp. pictured) is similar in appearance to a snake but it is distinguished by its broad fleshy tongue. This species has well developed hindlimb flaps and conspicuous external ear openings. They are widely distributed throughout most parts of continental Australia, except for the south-eastern coastal regions. Habitat is varied, from wet sclerophyll coastal forests and coastal heaths and dunes to semi-arid mallee. It is usually encountered sheltering under low vegetation, fallen timber and other ground litter. A shy lizard, it quietly slithers away when disturbed. It feeds upon insects and occasionally small skinks, and grows to over 60 centimetres in length.

One of the most formidable creatures in Australia, the estuarine or saltwater crocodile (*Crocodylus porosus*) is found in northern seas, estuaries and freshwater rivers and pools, from the Fitzroy River in the west, around the northern coastline and south down the eastern coast to about the Tropic of Capricorn. This large (up to 6 metres) and active predator is a very aggressive hunter, feeding on fish, waterfowl and animals. Large victims, such as kangaroos and wallabies, are usually seized by the head when they come to the water to drink. The crocodile then turns over and over in the water until the victim drowns. After making their kill they stash their prey under water, lodged in snags and river banks, returning to feed when the carcass begins to decompose. On occasion they become man eaters. These crocodiles hunt mostly at night and can often be seen during the day basking on river banks. Nests, made of decomposing vegetation and sand, are built on river banks and other flats just above high-tide mark. Females lay batches of up to sixty eggs.

(Below and right)
Averaging 1.2 metres in length, freshwater crocodiles (*Crocodylus johnstoni*) are smaller than the saltwater variety and are generally regarded as harmless. They are fairly widely distributed throughout the freshwater lagoons, rivers and billabongs of northern Australia, from the Kimberleys to Cape York. During the day they usually remain dormant, just floating in the water or lying under foliage. At night they forage for fish, frogs, birds, crustaceans, small mammals and reptiles. Just prior to the wet season females lay about twenty eggs, depositing them in hollows in sandbanks. A shy, unobtrusive animal, the freshwater crocodile usually submerges quietly at the slightest alarm.

Loggerhead turtles (*Caretta caretta*) are so named because of their big heads, which in old individuals can be almost 30 centimetres in length. Loggerheads have large brown eyes and pronounced beaks. The female comes ashore to lay about fifty eggs in a fairly shallow nest pit *(below)*. She then covers the eggs and heads back to the sea, using her flippers in a typically four-legged animal fashion *(left)*. Loggerheads are almost entirely carnivorous, feeding on crabs, molluscs, sponges, jellyfish and sometimes algae. They are found in the tropical and warm temperate waters off the coast, including the Great Barrier Reef. These pictures were taken at Mon Repos, a unique turtle rookery on the coast near Bundaberg, Queensland. Over a thousand loggerheads, as well as other breeds, lay eggs here from November to February.

Often called the blue fanny, the blue triangle (*Graphium sarpedon choredon*) is a very wide-ranging butterfly. It is found from southern coastal New South Wales to Cape York in Queensland. The sexes are similar in size and colouration; males have an average wingspan of 5.7 centimetres, females 6.3 centimetres. They lay their yellowish-green eggs on young shoots of foodplants, mainly camphor laurels, sassafras and laurel mahogany.

The wanderer or monarch butterfly (*Danaus plexippus*) is probably one of the best known butterflies in the world. Since its arrival around 1870 it has spread to all states except Western Australia. The males have a distinct sex mark—a small scent pouch on one of the veins of each hind wing. The perfume from this pouch is believed to attract the female. This fine butterfly has a strong rapid flight and is long-lived, specimens having being known to live for six months.

The larva of the wanderer butterfly is smooth creamy yellow with wide black transverse bands. There are a pair of long fleshy black tentacles at each end of the body. The wanderer's pupa is extremely beautiful: palest green decorated with delicate golden spots. It is suspended by the tail from a small pad of silk. The pupal stage is not long—about fourteen days.

One of Australia's largest and most common species of butterfly is the orchard (*Papilio aegeus aegeus*). Its range extends from Cape York down the eastern mainland to Victoria and South Australia. Citrus trees are food-plants of this butterfly and it is therefore often regarded as a pest. The male *(left)* is smaller than the female and has a totally different pattern of markings and colouration. The young caterpillar is greenish black with white markings and when fully grown *(below left)* is green with brown bands edged with white. The orchard butterfly flies rapidly and its wings are constantly in motion when it is feeding on flowers.

(Right)
The common grass yellow (*Eurema hecabe phoebus*) is a cheerful-looking little butterfly found from Sydney to northern Australia. The sexes are similar in size and colour, although some specimens lack the brown markings. They fly close to the ground, frequently land on flowers and can sometimes be seen drinking from puddles during hot weather. The eggs are laid on leaves of a small shrub (*Breynia oblongifolia*) and the fully grown larva is green with a pale yellow stripe on both sides. The elongated green pupa is very pointed at both ends.

(Right)
The Cairns birdwing (*Ornithoptera priamus*)—the largest butterfly in Australia, with the female attaining a wingspan of 19 centimetres. It ranges from Cooktown to Mackay in Queensland. The male is much smaller than the female (pictured) but far more colourful with brilliant green and black wings. In both sexes the thorax is marked with red. The very large round yellowish eggs are laid singly on several native species of Dutchman's pipe (*Aristolochia* spp.). The larva is dark brown with several longitudinal rows of pointed spines and a shiny black head with a white mark on the face. The large pupa is yellowish brown and has two laterally projecting spines on the head.

The common eggfly (*Hypolimnas bolina nerina*), one of the most common and most beautiful butterflies to be found along the southern, central and northern coastal regions. The females are extremely varied in colouration. The bluish-yellow eggs are laid on food-plants which include paddy's lucerne. The caterpillars, long and dark brown with rows of branching hairy spikes, feed at night and hide in sheltered spots during the day. The pupa is suspended head downwards by its tail from a pad of silk; it is dark brown and mottled and has several sharp spines along the abdomen.

One of the more unforgettable sounds of the Australian bushland is the song of the cicadas. They produce this communal chorus of 'drumming' by an elaborate pair of organs. The primary structures are parchment-like membranes set in deep pits on either side of the abdomen and capable of being vibrated by strong muscles. When distorted by the muscles the membranes emit a crackling sound, which when produced rapidly blends into the insect's ear-splitting song. The female lays her eggs in incisions in branches and twigs, and the eggs hatch into nymphs a few weeks later. They burrow into the soil and attach themselves to plant roots where they feed on sap for several years, until fully grown. They then climb the nearest tree, dig their feet firmly into the bark and shed their skins. During this metamorphosis the cicada is extremely vulnerable, so it instinctively ensures that the process is completed under the cover of darkness. By daylight, the insect is usually sheltering among the leaves. Cicadas have many predators (e.g., currawongs and sparrows), so their life is probably quite short—perhaps a few weeks. They are very ancient insects. The shales of Brookvale in New South Wales have yielded cicada fossils dating back to two hundred million years ago.

The Australian processionary caterpillars (*Ochrogaster contraria*) live in silken nests on eucalypts or wattle trees and emerge at night, when they range over the tree to feed on the leaves. While feeding, they spin a thread of silk, leaving a narrow ribbon to mark their journey. This is possibly used as a guideline to find their way back to shelter. Should the tree be too small to provide enough food, the insects move from tree to tree. They travel in a long line, each caterpillar with its head in contact with the tail of the caterpillar in front. When fully fed they march for the last time, searching for a suitable place to bury themselves under the soil, spin their cocoons and pupate.

One of Australia's best known moths—the emperor gum (*Antheraea eucalypti*). These large moths, with a wingspan of up to 12.5 centimetres, are often attracted by house lights and are well-known throughout eastern Australia. Emperor moths have no mouthparts and consequently cannot feed during their adult stage. These moths have conspicuous eyespots on the wings which are believed to be defence mechanisms as they are similar to the eyes of certain lizards and owls and thus deter potential predators. The female moth lays single eggs on the leaves of eucalypts. The extremely attractive caterpillar *(right)*, is large, green and fleshy, decorated with anemone-like tubercles or soft spikes on its back, tipped with blue or yellow. When fully grown, it settles down in one place and produces silk, gradually enclosing itself in a wall of fine threads. The threads are impregnated with a brown varnish-like liquid which hardens the cocoon. This water- and shockproof shelter is of the utmost importance because the pupal stage may last up to several years. The moth, at metamorphosis, has a small thorn at the base of the forewing to cut its way out of this extremely hard cocoon.

Golden orb-weavers (family Argiopidae) spin large webs of strong golden silk in forest areas and along electricity wires. There are eight species of orb-weavers, or *Nephila*, and they are found in the warmer states. They are large spiders with long legs that may be banded black and bright yellow. The web *(above)* is built vertically, strengthened on each side by a maze of strong supporting lines. The web's spiral is a true golden colour, but the supporting and radial lines of the orb are made from uncoloured silk. The web is not symmetrical, the hub usually being positioned at the top.

(Right)
The family Sparassidae (huntsman spiders or tri-antelopes) contains large hairy spiders with flat bodies which enable them to hide under the bark of trees. Huntsman spiders are generally regarded as harmless and the majority are timid creatures. In rainy weather they often come into houses where they scuttle sideways across walls and ceilings. The spider may be uniformly brown or grey, or have a pattern of darker mottling, or there may be a short longitudinal stripe on the otherwise plain abdomen.

Christmas or jewel spiders (*Gasteracantha minax*) are a common Australian species who prefer making their webs near stagnant water. The closely woven webs have a hole at the hub and are supported by a network of strong threads. The female spider is black, about 8 millimetres long, with a cream-coloured pattern on the back and orange legs with black bands. The male is much smaller—only 3 millimetres—and is black with a few lighter marks on the abdomen.

Some two hundred species of termites are known in Australia, many of which construct conspicuous mounds. Built to isolate the termite colony from the outside atmosphere, the mound gives the tiny creatures some control over their environment, especially humidity, during prolonged dry spells. Studies have shown that in the nursery area, in the heart of the termite mound, the relative humidity always exceeds 90 per cent. Much of this moisture comes from the metabolism of the insects themselves. The mounds vary greatly in size and shape—extremely large constructions being found in the northern parts of Australia, in regions which are hot and arid for long periods of the year. In Arnhem Land, the structures tower 6 metres or more with a 3.5 metre base. They are built of earth particles joined together with saliva, and when dry are as hard as concrete. A maze of channels, rooms and passages makes up the interior.

The deadly redback spider (*Latrodectus hasseltii*) is found across the whole continent, especially in the drier inland areas. The female spider has well-developed venom glands which can inject sufficient venom to prove fatal to an animal even as large as man. The male spider, by contrast, is believed to be harmless. The female, which may grow to 12 millimetres in length, has a large globular abdomen. There is a red hourglass design under the abdomen and usually a red longitudinal stripe on the back. Redbacks are found in sheltered places away from the wind and sun, under bushes, in heaps of dead wood, under empty tins in rubbish dumps, and in outbuildings.

INDEX

Numerals in *italics* denote photographs.

Acanthophis antarcticus 45
adders 15, *45*
agile wallaby (*Macropus agilis*) 23
amethystine python 15
amphibians 13
Amphibolurus barbatus 47
Amphibolurus decresii 47
Amphibolurus pictus 49
Antechinus stuartii 37
antechinuses 9, 37
Antheraea eucalypti 59
ants 16–17
Arctocephalus pusillus 12, 40, 41
Argiopidae 17, *60*
Australian fur seal (*Arctocephalus pusillus*) 12, 40, 41
Australian processionary caterpillar (*Ochrogaster contraria*) 58
Austrelaps superbus 43

bandicoots 11, 35
barking spider 17
bats 11, 41
bearded dragon (*Amphibolurus barbatus*) 47
bent-wings bats (*Miniopterus schreibersii*) 11, 41
Bettongia penicillata 27
bettongs 11, 27
bilbies 11
birdwing butterflies 17, 56, 57
black snake, red-bellied (*Pseudechis porphyriacus*) 44
blind snakes 15
'blue flyer' *see* red kangaroo
blue-tongued lizard (*Tiliqua scincoides*) 14, *50*
blue triangle butterfly (*Graphium sarpedon choredon*) 55
Boidae 15, 44, *44*
brown antechinus (*Antechinus stuartii*) 37
brown tree snake 15
brush-tailed bettong (*Bettongia penicillata*) 27
brush-tailed phascogale (*Phascogale tapoatafa*) 37
brush-tailed possum (*Trichosurus vulpecula*) 32, *33*
brush-tailed rock-wallaby (*Petrogale pencillata*) 25
bush flies 16, 17
butterflies 17, 55–7

Cairns birdwing butterfly (*Ornithoptera priamus*) 56, 57
cane toad 13
Canis familiaris 7, 12, 38, *39*
Caretta caretta 14, *54*
carpet pythons 15
caterpillars 58, *59*
Cercartetus nanus 34
Christmas spider (*Gasteracantha minax*) 17, *62*
cicadas 16, *58*
colubrine snakes 15
common death adder (*Acanthophis antarcticus*) 45
common eggfly butterfly (*Hypolimnas bolina nerina*) 56
common garden spiders 17
common grass blue butterflies 17
common grass yellow butterfly (*Eurema hecabe phoebus*) 56, 57
common wombat (*Vombatus ursinus*) 28
copperhead snakes 15, 43
crab-eater seals 12
crab spiders 17
crest-tailed marsupial rat (*Dasyuroides byrnei*) 35
crocodiles 14, *14*, 17, 52–3
Crocodylus johnstoni 14, *14*, *52*, *53*
Crocodylus porosus 14, *52*

Danaus plexippus 55
Dasyuroides byrnei 35
Dasyurops maculatus 36
Dasyurus viverrinus 36

Delias argenthona 17
dingo (*Canis familiaris*) 7, 12, 38, *39*
dragon lizards 14, 47–9
dugongs 12
dunnarts 9, 36

eastern brown snakes 15
eastern grey kangaroo (*Macropus giganteus*) 21
eastern pygmy possum (*Cercartetus nanus*) 34
eastern quoll (*Dasyurus viverrinus*) 36
easter tiger snake (*Notechis scutatus*) 43
eastern water dragon (*Physignathus lesueurii*) 48
echidna (*Tachyglossus aculeatus*) 8, 12, 42
emperor gum moth (*Antheraea eucalypti*) 59
estuarine crocodile (*Crocodylus porosus*) 14, *52*
Eurema hecabe phoebus 56, 57
euros 10, 11

flat-back turtles 14
flies 16, 17
flower spiders 17
flying foxes 11
forester kangaroos *see* grey kangaroos
freshwater crocodile (*Crocodylus johnstoni*) 14, *14*, *52*, *53*
frill-necked lizards 14
frogs 13
fruit-bats 11
funnel-web spiders 17

Gasteracantha minax 17, *62*
geckos 15, 46
ghost bats 11
goannas 14–15, 49–50
golden orb-weaver spider (*Argiopidae* family) 17, *60*
Graphium sarpedon choredon 55
great grey kangaroos *see* grey kangaroos
green tree snakes 15
green turtles 14
grey kangaroos 7, 11, 21, 22

hairy-nosed wombat (*Lasiorhinus latifrons*) 28
hawksbill turtle 14
honey possum 9
hopping-mice 12, 38
horseshoe bats 11
huntsman spider (*Sparassidae* family) 17, *60*, *61*
Hypolimnas bolina nerina 56

insects 16–17, 55–9
Isoodon obesulus 35

jewel spider (*Gasteracantha minax*) 17, *62*
jumping spiders 17

Kangaroo Island kangaroo (*Macropus fuliginosus fuliginosus*) 18
kangaroos 7, 9, 10, 11, 12, 18–22
keelback snakes 15
knob-tailed gecko (*Nephrurus levis*) 46
koalas (*Phascolarctos cinereus*) 9, 17, 28, 29–31
kowari (*Dasyuroides byrnei*) 35

lace monitor (*Varanus varius*) 49
Lasiorhinus latifrons 28
Latrodectus hasseltii 17, *62*
legless lizard (*Delma* sp.) 15, *51*
leopard seals 12
Lesueur's gecko (*Oedura lesueurii*) 46
Liasis amethystinus 44
Liasis fuscus 44
lizards 14–15, 46–51
loggerhead turtle (*Caretta caretta*) 14, *54*
long-beaked echidna 12
long-eared bats 11
long neck tortoises 14
long-nosed potoroo (*Potorous tridactylus*) 27
long-tailed spiny skinks 14
lowland copperhead snake (*Austrelaps superbus*) 43

Macleay's water snake 15
macropods 7, 9, 11, 12, 18–27
Macropus agilis 23
Macropus eugenii 25

Macropus fuliginosus fuliginosus 18
Macropus giganteus 21
Macropus parma 25
Macropus parryi 22
Macropus rufogriseus 23
Macropus rufus 10, 11, 18–20, 21
mainland tiger snake (*Notechis scutatus*) 43
mammals 7–12, 18–42
marsupial mice 8, 11, 37
 see also dunnarts
marsupial moles 8
marsupial rats 35
marsupials 7, *7*, 8–11, 12, 18–37
mastiff bats 11
Merten's water monitor (*Varanus mertensi*) 50
Miniopterus schreibersii 11, 41
Moloch horridus 14, *50*
monarch butterfly (*Danaus plexippus*) 55
monitor lizards 14–15, 49–50
monotremes 8, 12, 42
mosquitos 17
moths 59
mulga snakes 15
musky rat-kangaroo 11

Nephrurus levis 46
New Zealand fur seals 12
Notechis scutatus 43
Notomys alexis 38
numbats 9

Ochrogaster contraria 58
Oedura lesueurii 46
Ogyris 17
orchard butterfly (*Papilio aegeus*) 56
Ornithoptera priamus 56, 57
Ornithorhynchus anatinus 8, 12, 17, 42
Oxyuranus scutellatus 6, 15, 45

pademelons 11, 26–7
painted dragon (*Amphibolurus pictus*) 49
Papilio aegus 56
parma wallaby (*Macropus parma*) 25
Petaurus breviceps 32
Petaurus norfolcensis 32
Petrogale penicillata 25
Petrogale xanthopus 10, *26*
Phascogale tapoatafa 37
phascogales 9, 37
Phascolarctos cinereus 9, 17, 28, 29–31
Physignathus lesueurii 48
placental mammals 7, 8, 11–12, 38–41
plains kangaroos *see* red kangaroo
planigales 9
platypus, duck-billed (*Ornithorhynchus anatinus*) 8, 12, 17, 42
possums 8, 9, 17, 32–4
potoroos 11, 27
Potorous tridactylus 27
Pseudechis porphyriacus 44
Pygmy possum, eastern (*Cercartetus nanus*) 34
pygopods 15, *51*
pythons 15, 44, *44*

Queensland rock python (*Liasis amethystinus*) 44
quokka (*Setonix brachyurus*) 11, 27
quolls 8, 9, 36

rat-kangaroos 11
Rattus lutreolus 38
red-bellied black snake (*Pseudechis porphyriacus*) 44
red kangaroo (*Macropus rufus*) 10, 11, 18–20, 21
red-legged pademelon (*Thylogale stigmatica*) 26
red-necked pademelon (*Thylogale thetis*) 26
red-necked wallaby (*Macropus rufogriseus*) 23
redback spider (*Latrodectus hasseltii*) 17, *62*
reptiles 13, 14–15, 43–51
rock-rats 12
rock-wallabies 10, 11, 25, *26*
rodents 11–12, 38

saltwater crocodile (*Crocodylus porosus*) 14, *52*
Sarcophilus harrisii 9, 35
sea cow *see* dugongs
sea lions 12

seals 12, 17, 40, 41
Setonix brachyurus 11, 27
sheath-tailed bats 11
shingleback lizard (*Trachydosaurus rugosus*) *51*
short-beaked echidna (*Tachyglossus aculeatus*) 8, 12, 42
short neck tortoises 14
skinks 14
Sminthopsis macroura 36
snake neck tortoises 14
snakes 15, 43–5
southern brown bandicoot (*Isoodon obesulus*) 35
southern elephant seals 12
southern frogs 13
Sparassidae 17, *60*, *61*
spiders 17, *60*–2
spinifex hopping-mice (*Notomys alexis*) 38
spiny anteater (*Tachyglossus aculeatus*) 8, 12, 42
spotted cuscus 9
spotted-tailed quoll (*Dasyurops maculatus*) 36
squirrel glider (*Petaurus norfolcensis*) 32
stick-nest rats 12
stripe-faced dunnart (*Sminthopsis macroura*) 36
sugar glider (*Petaurus breviceps*) 32
swamp rat (*Rattus lutreolus*) 38
swamp wallaby (*Wallabia bicolor*) 24

Tachyglossus aculeatus 8, 12, 42
taipan (*Oxyuranus scutellatus*) 6, 15, 45
tammar wallaby (*Macropus eugenii*) 25
Tasmanian devil (*Sarcophilus harrisii*) 9, 35
Tasmanian tiger 9
tawny dragon (*Amphibolurus decresii*) 47
tent spiders 17
termites 16, 17, *62*
thorny devil lizard (*Moloch horridus*) 14, *50*
Thylacoleo 7
Thylogale stigmatica 26
Thylogale thetis 26
tiger cats *see* quolls
tiger snakes 15, 43
Tiliqua scincoides 14, *50*
tortoises 14
Trachydosaurus rugosus *51*
trapdoor spiders 17
tree frogs (*Hylidae*) 13
tree-kangaroos 11
tree-rats 12
tree snakes 15
Trichosurus vulpecula 32, *33*
turtles 14, 17, *54*
Typhlopidae 15

Ulysses butterflies 17

Varanus mertensi 50
Varanus varius 49
Vombatus ursinus 28

Wallabia bicolor 24
wallabies 10, 11, 22, 23–7
wallaroos *see* euros
wanderer butterfly (*Danaus plexippus*) 55
water-holding frogs 13
water python (*Liasis fuscus*) 44
water-rat 12, 17
Weddell seals 12
western grey kangaroos 18
whiptail wallaby (*Macropus parryi*) 22
white cabbage butterflies 17
wolf spiders 17
wombats 9, 28
wood frogs 13

yellow-footed rock-wallaby (*Petrogale xanthopus*) 10, *26*